WORKING WITH MATERIALS

Sonya Newland

Kane Miller
A DIVISION OF EDC PUBLISHING

First American Edition 2022
Kane Miller, A Division of EDC Publishing

Editor: Sonya Newland
Illustrator: Diego Vaisberg
Designer: Clare Nicholas

First published in Great Britain in 2020 by Wayland, An imprint of Hachette Children's Group, Part of The Watts Publishing Group, Carmelite House, 50 Victoria Embankment, London EC4Y 0DZ

For information contact:
Kane Miller, A Division of EDC Publishing
5402 S 122nd E Ave, Tulsa, OK 74146
www.kanemiller.com
www.myubam.com

Library of Congress Control Number: 2021937020

Printed and bound in China
1 2 3 4 5 6 7 8 9 10

ISBN: 978-1-68464-332-5

All the materials required for the projects in this book are available online, or from craft or hardware stores. Adult supervision should be provided when working on these projects.

CONTENTS

WHAT ARE MATERIALS?

Look around you. Everything you can see is made of materials. The chair you're sitting on, the TV you're watching, the clothes you're wearing ... All these objects have been created by amazing engineers.

Types of materials

Engineers design all sorts of things, from socks to skyscrapers. When they start to tackle a task, one of the first questions they ask is: "What should this object be made of?" They can choose from many different materials.

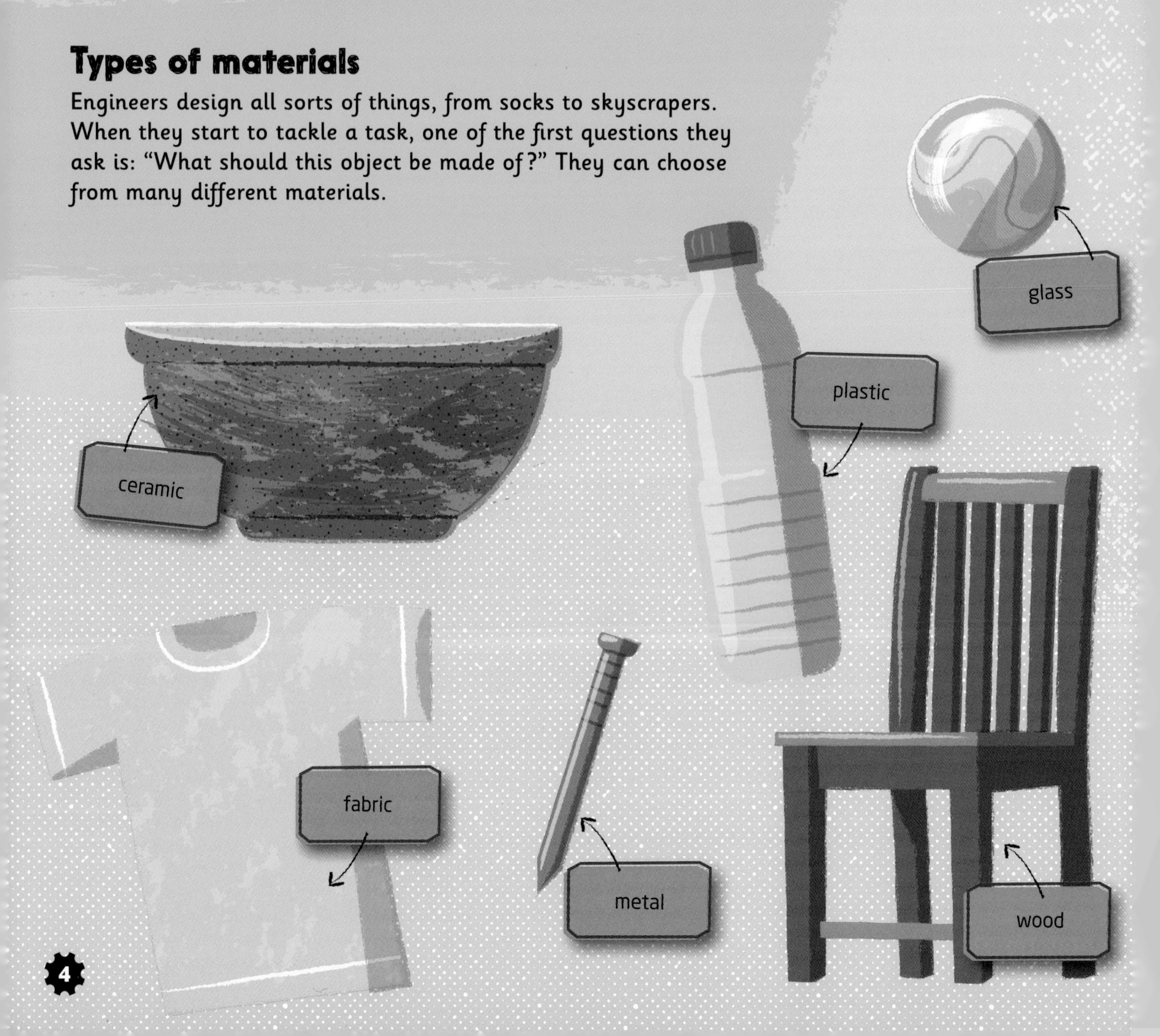

Natural or man-made?

Natural materials include organic substances, which come from living things, such as wood and wool, and inorganic materials such as rocks, minerals, and some metals. Man-made materials are ones that engineers design in laboratories and create in factories. They include plastic and glass.

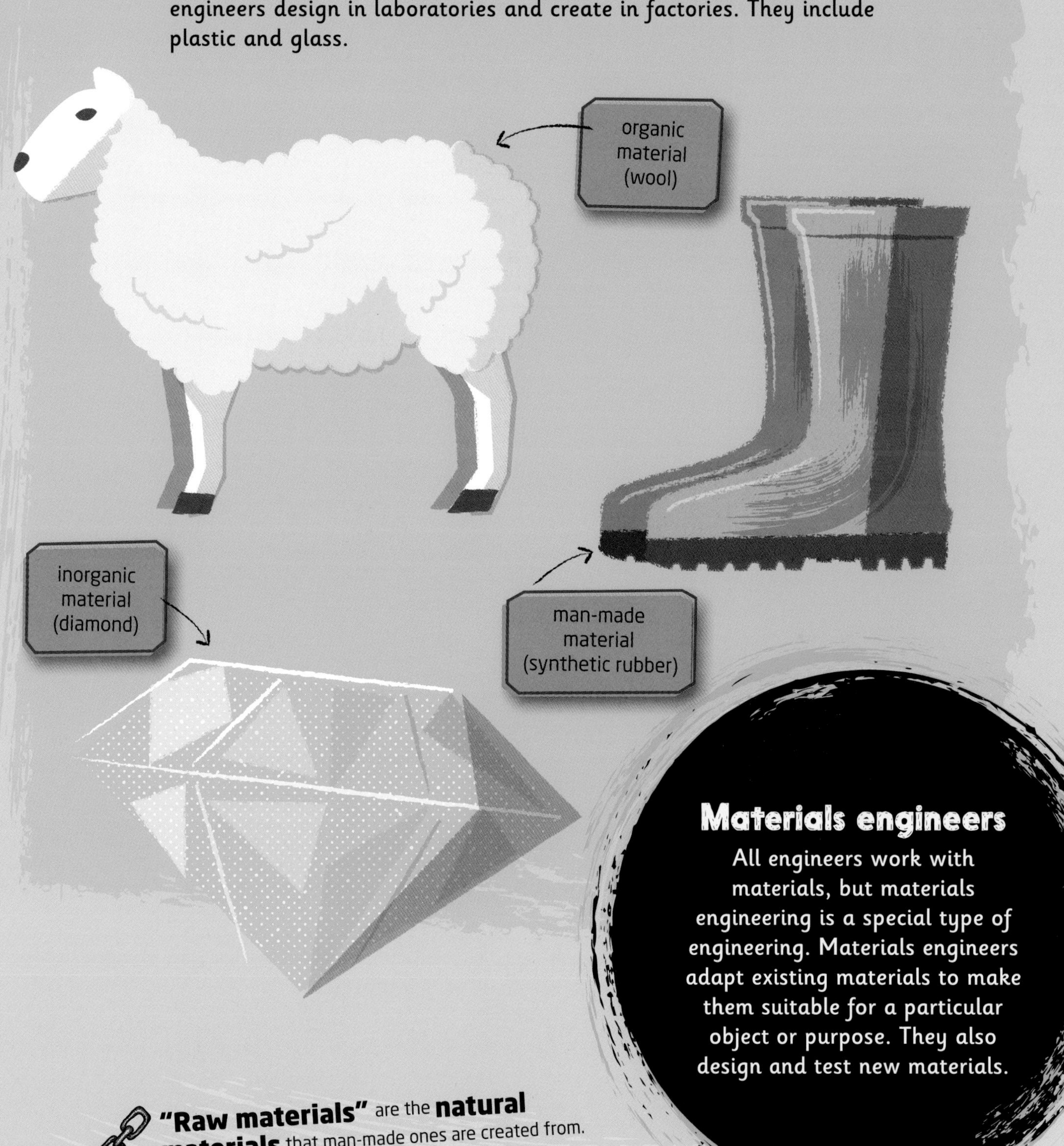

Materials engineers

All engineers work with materials, but materials engineering is a special type of engineering. Materials engineers adapt existing materials to make them suitable for a particular object or purpose. They also design and test new materials.

"Raw materials" are the **natural materials** that man-made ones are created from. For example, **iron and carbon** are the raw materials used to make **steel.**

PROPERTIES OF MATERIALS

Different materials have different properties. Their properties are what help engineers choose the right material for the job.

Materials for every purpose

There's no such thing as a good or bad property in a material.
It all depends on what you want to use it for. For each property shown, ask yourself:

- What advantages can you think of for this property?
- What uses do you think this property makes the material suitable for?

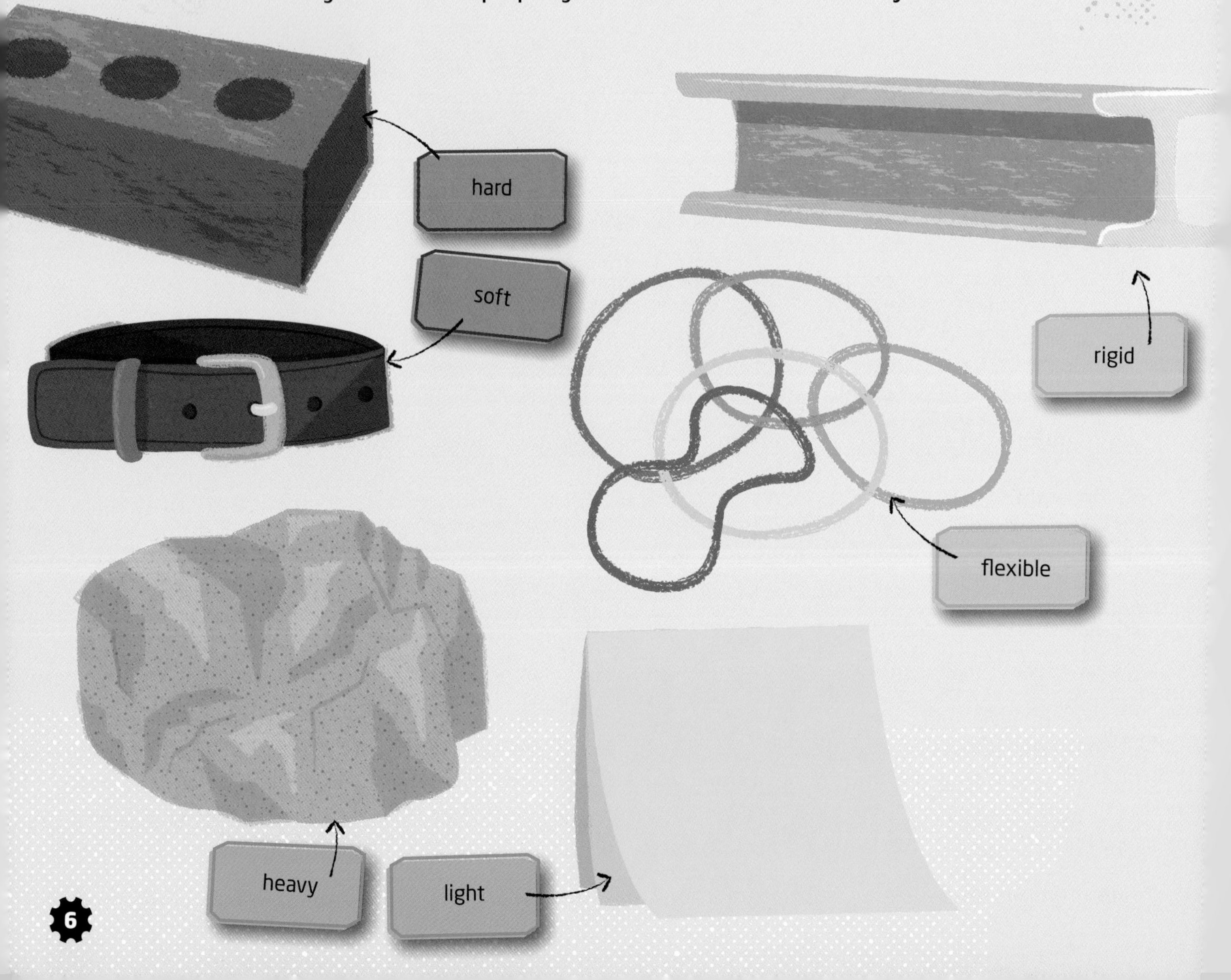

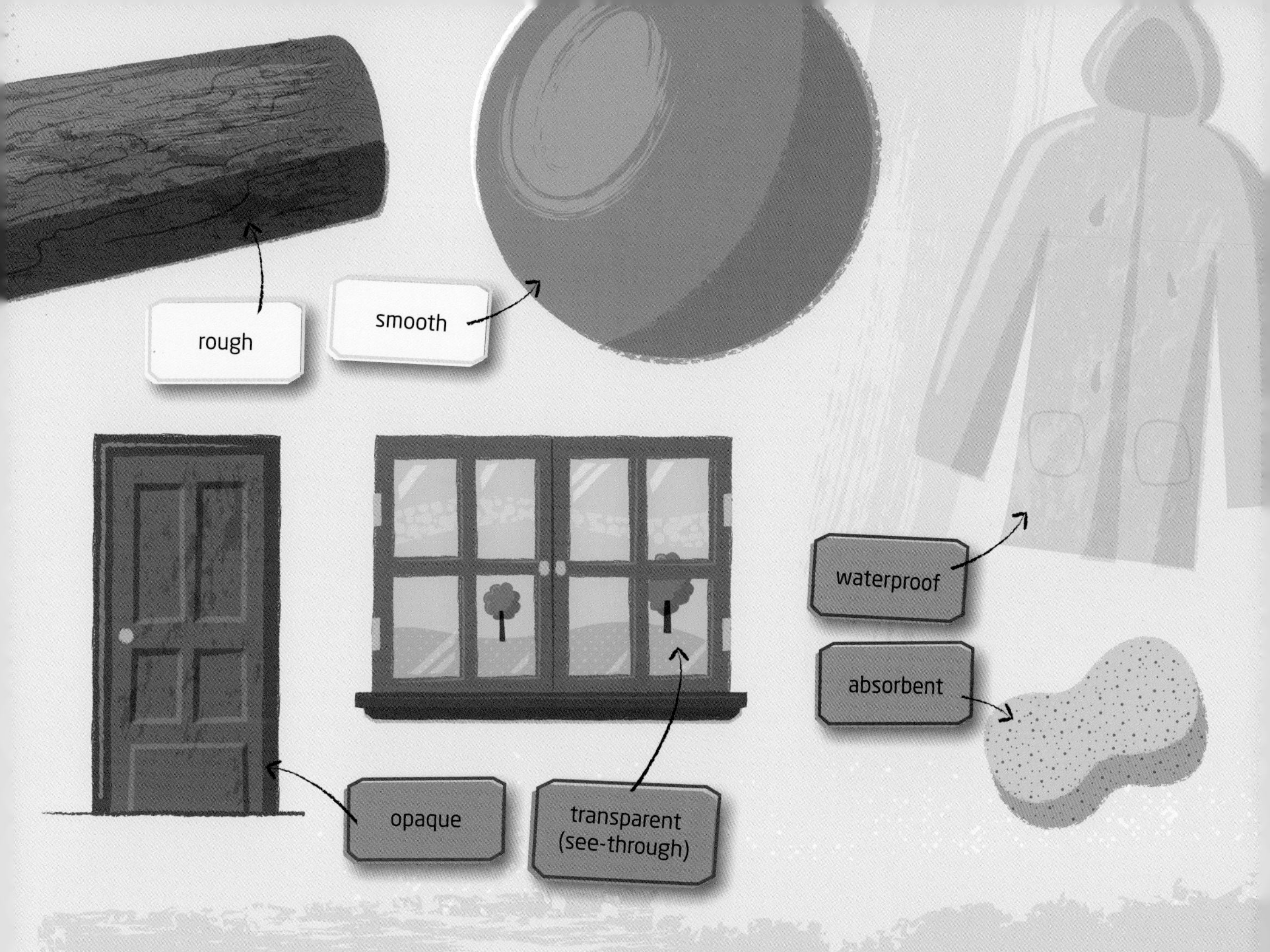

Mix and match

Of course, materials don't have just one property.

Glass can be ...

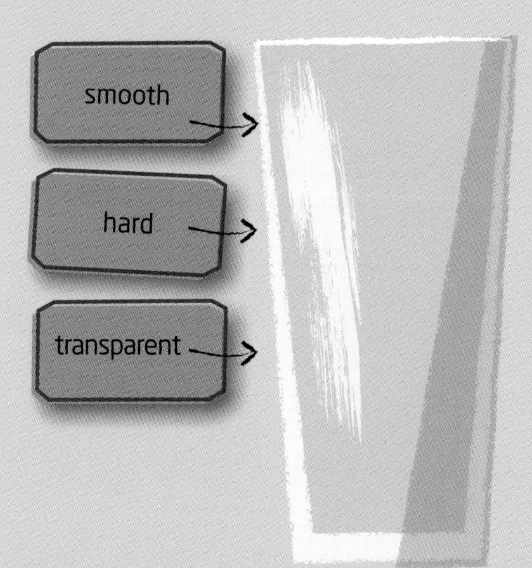

Paper can be ...

Engineers can **change** a material's **properties.** For example, clay is **soft and flexible,** but when it is fired it becomes **hard ceramic.**

YOU'RE THE ENGINEER: TESTING PROPERTIES

Understanding the basic properties of a selection of materials.

You will need

Three rulers of the same size made of:
- wood
- plastic
- metal

A piece of cardboard the same size as the rulers
Masking tape
A notepad and pencil
A length of string
Two 100 g weights

1 Place the three rulers and piece of cardboard on the edge of a table or desk so that half of each is on the table and half hanging over the edge.

2 Fix each of them securely in place with the masking tape.

3 In your notepad, draw a table with four columns. Write the name of the material each ruler is made of in the first column. In the second column, write down what properties you think the material has, such as "rigid" or "flexible." In the third column, predict what will happen when you apply weight to the material.

Material	Predicted properties	Predicted behavior	Actual properties
Wood			
Plastic			
Metal			
Cardboard			

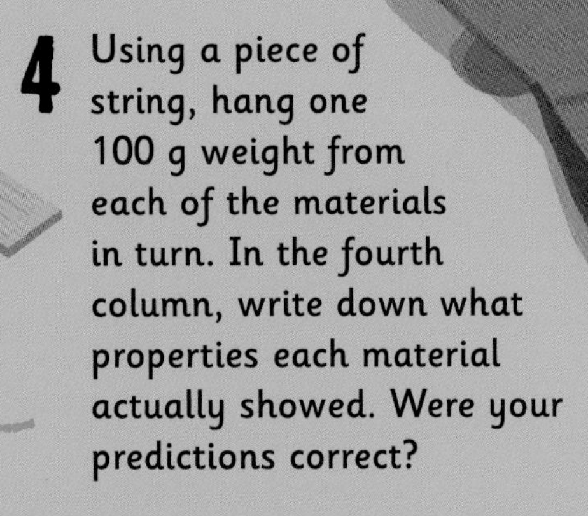

4 Using a piece of string, hang one 100 g weight from each of the materials in turn. In the fourth column, write down what properties each material actually showed. Were your predictions correct?

5 Now try doubling the weight on each material, using both 100 g weights. Compare the effect. Does the plastic bend twice as far, for example?

TEST IT!

Vary the experiment in different ways. For example, how might you change certain properties? Think about how you could make each ruler stronger or more rigid. Each time, write down your findings and compare them.

FRICTION

One of the key things engineers have to consider when choosing which materials to use is friction. Friction is the force created when two surfaces rub together.

Surface friction

Different surfaces create different amounts of friction.

Rough:
Materials with rough surfaces, such as sandpaper or gravel, create a lot of friction.

Smooth:
Materials with smooth surfaces, such as ice or polished metal, create much less friction.

Friction everywhere

Even air and water create friction, slowing down planes and boats. Engineers who design these types of crafts consider friction in two main ways.

They design the craft in slim, pointy, streamlined shapes that allow air or water to flow past them more easily. This causes less friction.

They choose materials that create low surface friction, such as the smooth, flexible metal aluminum.

Changing friction

Sometimes engineers want to reduce friction, to make an object move more smoothly. At other times they want to increase friction, to make something move more slowly. Sometimes they need to do both, such as when designing different parts of a car.

Car bodies are usually made of smooth steel, which reduces friction between the air and the vehicle.

Oil reduces friction between touching surfaces so that parts of the engine move smoothly.

Friction between the brakes and the wheels is what slows a car down.

Tires are made of rubber, which grips the road and stops the car from skidding.

Tread (a pattern of grooves) in the tires creates a rougher surface, which increases friction and improves grip.

GOT IT!

Engineers at the University of Wisconsin-Madison are developing a tiny generator that may one day be built into cars. This amazing device could turn the energy created by the friction between the tires and the ground into a power source for the car!

YOU'RE THE ENGINEER:
RACETRACK FRICTION

Find out which surfaces create the most friction by testing four ramps with different surfaces.

You will need

- Four planks of wood the same size
- A towel
- Sandpaper
- Aluminum foil
- Books
- A toy car
- A stopwatch
- A notepad and pencil

1 Create your ramps by covering three of the planks in the towel, the sandpaper, and the aluminum foil. Do not cover the fourth plank. What are the properties of each material? Which one do you think will create the most friction? Which one the least?

2 Now, engineer your ramps. How high will you make them? Find a suitable surface to support the ramps, such as a low bench. If you want a steeper angle, pile books on the bench to make it higher. Then prop your ramps up against them.

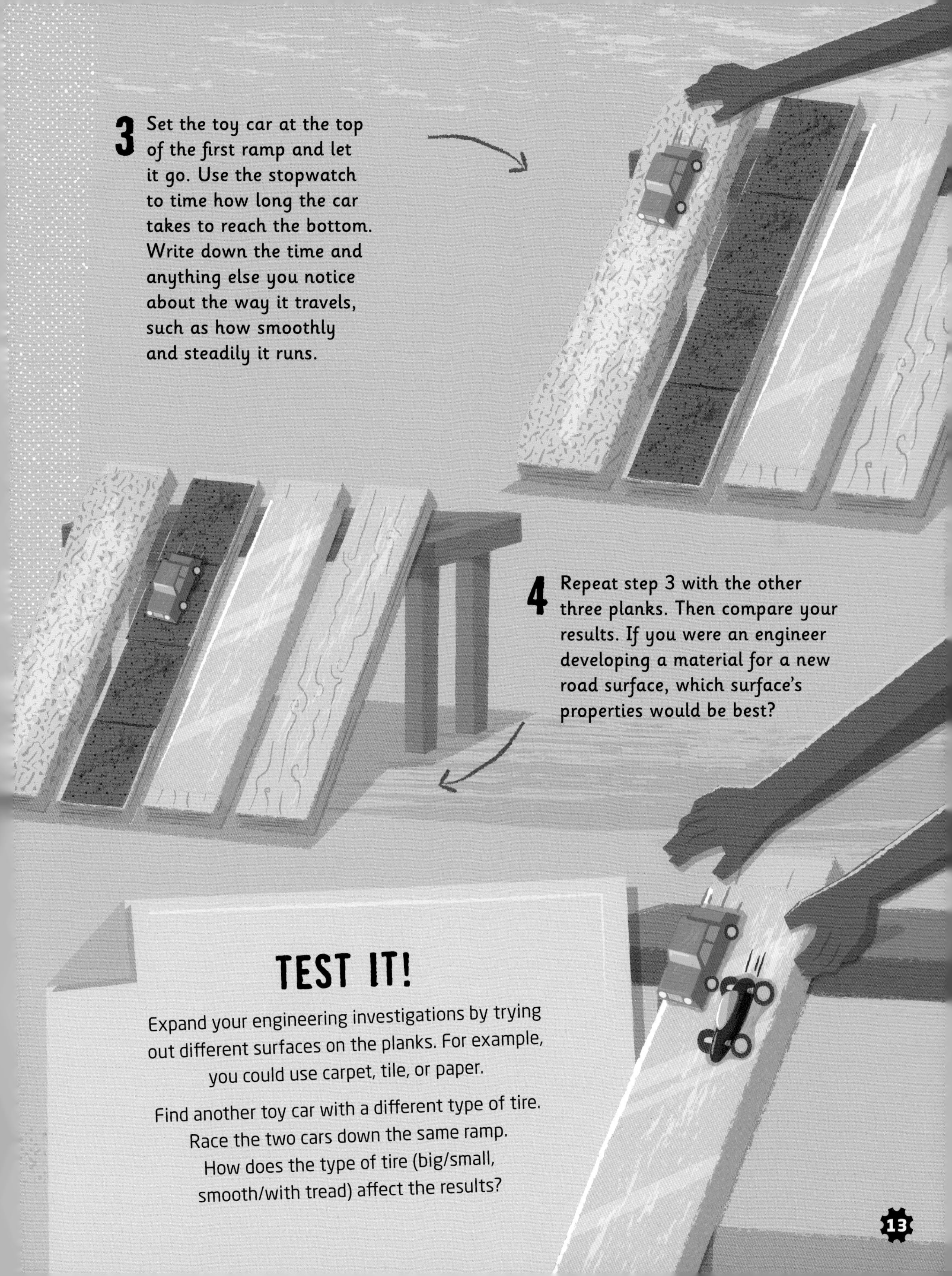

3 Set the toy car at the top of the first ramp and let it go. Use the stopwatch to time how long the car takes to reach the bottom. Write down the time and anything else you notice about the way it travels, such as how smoothly and steadily it runs.

4 Repeat step 3 with the other three planks. Then compare your results. If you were an engineer developing a material for a new road surface, which surface's properties would be best?

TEST IT!

Expand your engineering investigations by trying out different surfaces on the planks. For example, you could use carpet, tile, or paper.

Find another toy car with a different type of tire. Race the two cars down the same ramp. How does the type of tire (big/small, smooth/with tread) affect the results?

METAL

Some metals occur naturally in the earth, such as iron, copper, nickel, and titanium. Others are man-made – the result of engineers mixing natural metals and minerals. For example:

iron + carbon = steel
copper + nickel = cupronickel

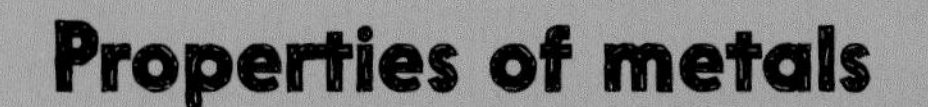

Properties of metals

They are good conductors of heat and electricity.

They can be hammered into different shapes without breaking.

They are hard, strong, and usually shiny.

Attractive metals

Some metals are magnetic – but not all!

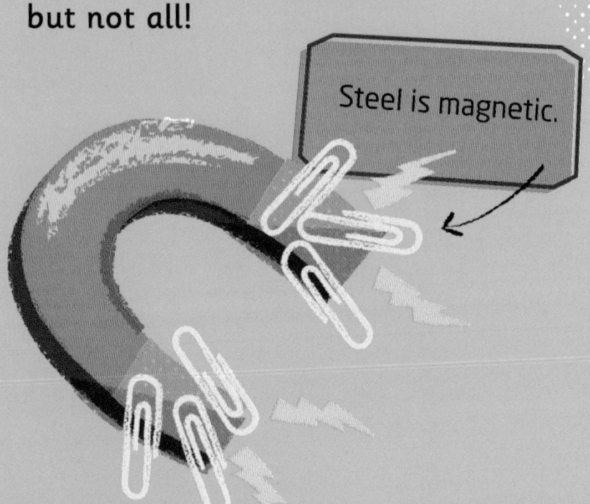

Steel is magnetic.

Aluminum is not magnetic.

Cool construction

Civil engineers, who design and build buildings and bridges, would be out of a job without metals! All these structures are built around frames of incredibly strong metals such as iron and steel.

Mining metals

Many raw materials, including metals, have to be dug from the earth before people can put them to use. That's where mining engineers come in. They work out ways to dig out the metal ore from the earth, and then decide how to extract the useful metal from the ore.

PLASTIC

Plastics are big business for materials engineers. Plastic can be found in millions of different products – from pens to clothes to cars. There's even plastic in some countries' banknotes used to buy these items!

Fantastic plastic

Plastic is an incredibly useful material. It is light, strong, and versatile. It is also cheap and easy to manufacture. There are more than 50 different types of plastic, but they all belong to one of two main sets.

Thermoplastics become soft and malleable (easily molded) when heated, but harden into strong plastic when they cool, like these toy blocks. They can be melted and remolded several times.

Thermoset plastics are made straight into solid products. Once hardened, thermoset plastic cannot be melted again. Why do you think electrical items like this iron are made of thermoset plastics?

A plastic problem

Unfortunately, the properties that make plastic so useful also cause problems. Because it is so long-lasting, plastic waste has to be buried in landfills. Some plastic can be recycled, but in landfills, plastic can take hundreds of years to biodegrade.

New plastics

Materials engineers are working hard to solve the problem of plastic pollution by creating new plastics that can be recycled more easily or that will biodegrade like organic materials (see page 5).

GOT IT!

Plastic was invented in 1907 by a Belgian chemist called Leo Baekeland (1863–1944). He wanted a strong, moldable material that would not conduct heat or electricity. He called his product Bakelite, and it was soon being used to make all sorts of products, from telephones to toys!

YOU'RE THE ENGINEER:
CREATE A NEW PLASTIC

Find out how materials engineers change the properties of plastic with some help from science!

You will need

PVA glue (white glue)
A small glass bowl
A tablespoon
A notepad and pencil
Borax solution
Food coloring
Rubber gloves

1 PVA glue is a type of plastic. Begin by putting two tablespoons of glue into the bowl. In your notepad, write down what properties it has, such as "soft," "runny," "sticky."

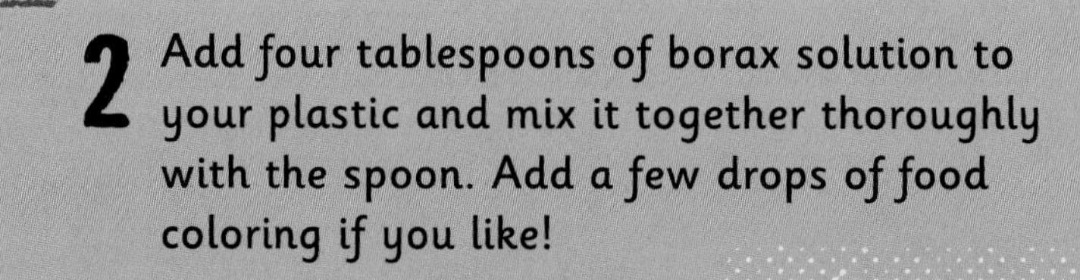

2 Add four tablespoons of borax solution to your plastic and mix it together thoroughly with the spoon. Add a few drops of food coloring if you like!

The **raw material** that plastics are made from is **crude oil.** This **fossil fuel** is a dark liquid found **deep below ground.**

3 Put on the rubber gloves and knead the plastic mixture in the bowl until it is firm and rubbery. Mold it into a ball.

4 Remove the ball from the bowl and wash it carefully in clean water.

5 Experiment to discover what properties you can see and feel now. Is the ball solid? Does it bounce? Write down what changes have occurred to the plastic mixture.

TEST IT!

PVA glue is a type of plastic made up of long "threads" that are not linked. The borax solution connects the threads to make the plastic more solid and rubbery. What happens if you change the amounts of glue and solution? Put the ball back in the bowl and leave it for a few hours. What happens to the mixture?

GLASS AND CERAMICS

Glass and ceramics are both man-made materials. You might think that, since they break easily, they are not very useful, but engineers can adapt them for thousands of different uses.

Great glass

Glass is colorless, transparent, and waterproof. It is strong and long-lasting if care is taken with it. But it is also brittle, which means it breaks easily.

When glass is molten, it can be molded into any shape.

Colors can be added to make glass items in every color of the rainbow.

Engineered glass

Materials engineers add different ingredients to glass to change its properties. Putting layers of plastic in between layers of glass makes it so strong that it's bulletproof!

Fiberglass is long strips of glass layered or woven together. It is very strong and light, so it is used to make things like aircraft and surfboards.

GOT IT!

In the early 1900s, engineers developed a new kind of glass. Its official name is borosilicate, but it's commonly known as Pyrex®. This special glass can withstand high temperatures. It is thicker and stronger than normal glass, so it is less likely to break. These properties make it perfect for kitchenware!

Amazing clay

Ceramics are man-made materials produced from minerals called clay. Engineers might use clay if they need a material that is heat resistant and durable. Ceramics include:

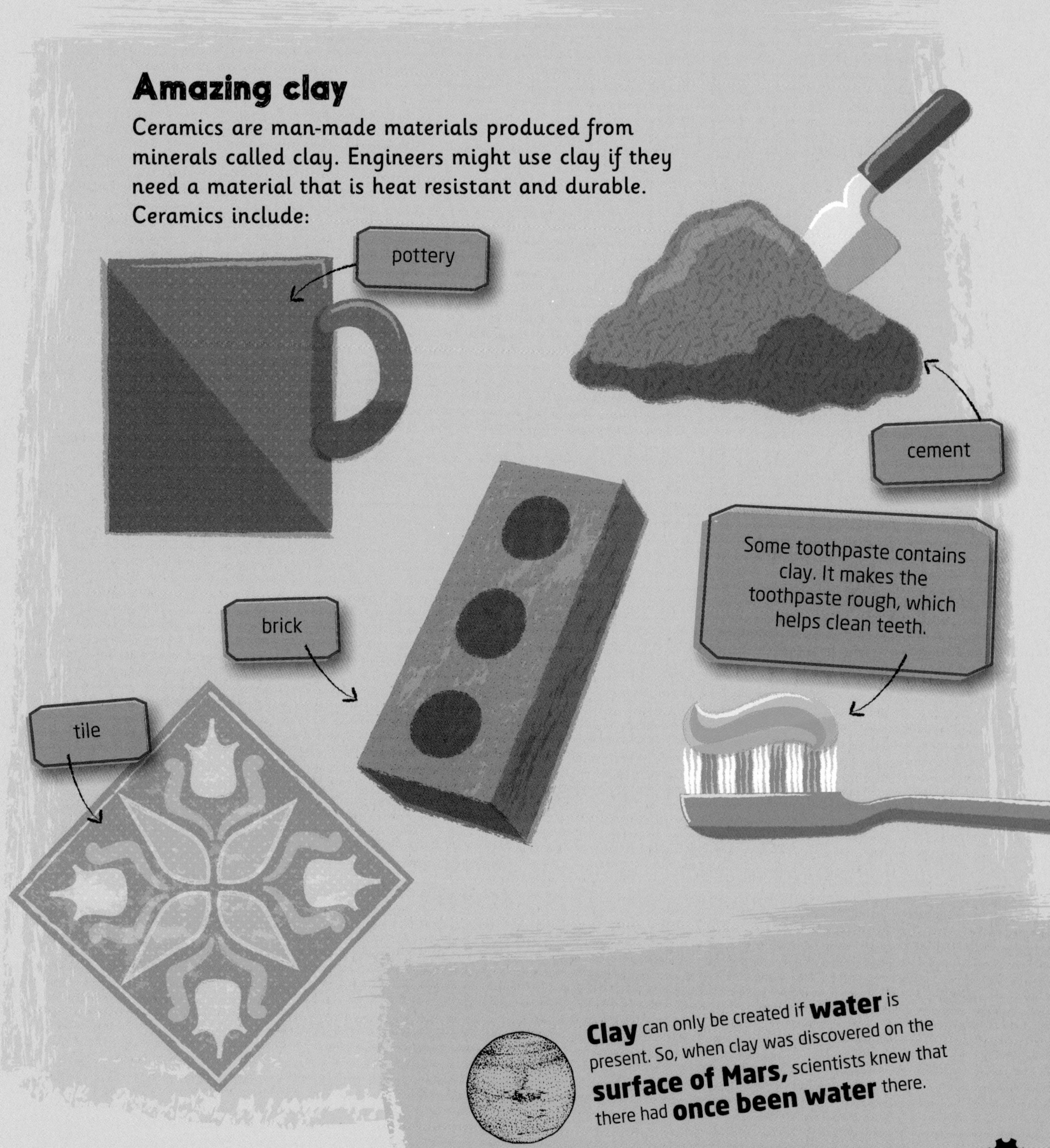

Clay can only be created if **water** is present. So, when clay was discovered on the **surface of Mars,** scientists knew that there had **once been water** there.

YOU'RE THE ENGINEER:
CHANGING CLAY

Depending on how much water is applied, clay can take on the properties of other materials – from leather to stone. See for yourself!

You will need

- A piece of glazed pottery
- A notepad and pencil
- A piece of pottery before it has been glazed
- A lump of clay, fully dried out
- A plastic bowl
- Water
- Newspaper

1 Look carefully at your glazed pottery piece. Glazed pottery has been fired in an oven called a kiln, then painted and fired a second time. What properties can you see and feel? How strong do you think this type of ceramic is? Could it become wet clay again? Write down your thoughts in your notepad.

2 Bisqueware is pottery that has been fired once. Look at your unglazed piece of pottery. How hard do you think it is? What other properties does it have? Could it be turned back into wet clay? Rub some water on the outside and see! Jot down your findings.

3 Now take the lump of dry clay and put it in the bowl. This stage is called "bone dry." Use your hands to test its properties. What does it feel like? Does it crumble easily? Add a tiny bit of water. What do you notice?

4 Add a few spoonfuls of water, one at a time, and mix each one in with your hands until the clay is flexible but still slightly dry. Look back at the pictures showing the properties of materials on pages 6–7. Which material does the clay's consistency most feel like? (Check your answer on page 31.)

5 Add a few more spoonfuls of water, one at a time, until the clay feels sticky and can be molded easily. This is called the "plastic" stage. Take it out and put it on a piece of newspaper. Can you mold it into the same shape as the pot in step 1? Write a list of properties that the clay has now.

TEST IT!

When you have finished shaping the clay, put it back into the bowl and add more water until the clay has the same properties as the PVA glue you used on pages 18–19. This is called "slip." Potters use this as a glue to stick together pieces of clay in other forms.

WOOD AND TEXTILES

Plants are the source of two important materials that we use all the time. Trees are cut down for wood, and the leaves and fibers of plants are used to create textiles.

Wonderful wood

Wood is strong, but also flexible when cut in certain ways. It is long-lasting and able to be carved into different shapes. It does not conduct heat or electricity. All these properties mean engineers can use wood to design lots of different objects, from golf tees to garden furniture!

Trees and plants are essential to our planet. To make sure we always have enough trees, we need to try to plant at least as many as we cut down.

Textiles

Textiles are materials that are made by weaving or knitting fibers together. Engineers can use plant fibers such as cotton, flax, and raffia. They can also use animal fibers such as wool, cashmere, and angora.

Modern textiles

Not all fabrics are natural. Engineers have created many new types of textiles. Most of these are polymers. They are made by forcing liquid chemicals through tiny holes, which dries them into threads. Polymers are designed to have special properties. For example, the polymer polyester does not wrinkle so does not need ironing; acrylic is warm and soft.

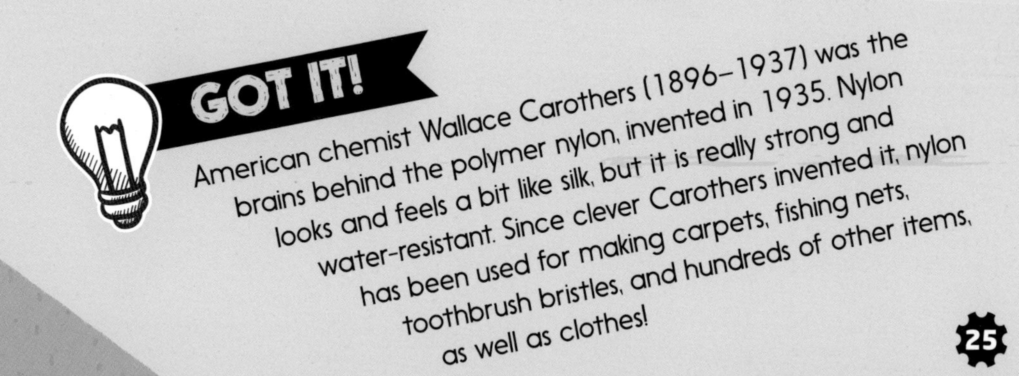

GOT IT!

American chemist Wallace Carothers (1896–1937) was the brains behind the polymer nylon, invented in 1935. Nylon looks and feels a bit like silk, but it is really strong and water-resistant. Since clever Carothers invented it, nylon has been used for making carpets, fishing nets, toothbrush bristles, and hundreds of other items, as well as clothes!

MATERIALS OF THE FUTURE

Engineers are always developing materials that are stronger, lighter, or more flexible. They are also aware that we need to reduce our use of natural resources, such as wood, and make whatever products we can more easily recyclable.

Understanding atoms

Nanotechnology is the secret to inventing stronger, lighter, cutting-edge materials. This is science on a tiny scale. All materials are made up of tiny particles called atoms. There are 118 types of atom, or elements, which are the building blocks of all materials. Engineers create new materials by changing the way that atoms are arranged in existing materials.

New materials might be used for all sorts of everyday objects ...

Different structures, different properties

Tiny changes can make a big difference to a material. Carbon is a chemical element, but it comes in different forms depending on how it is arranged on a nanoscale.

Diamond is carbon atoms arranged in rigid layers. Diamond is incredibly strong.

Fullerenes are carbon atoms in balls or tubes. They are light but strong, so are used in objects such as bicycle frames.

Graphite is carbon atoms arranged in sheets that can move over each other. This makes it quite soft and slippery, so it is useful for writing with.

Amazing materials

In recent years, engineers have created some truly amazing materials. Aerogel is a kind of "frozen smoke." It is light, as well as being an amazing insulator. It can be used to insulate windows and thicken paint. A form of alumina is three times as strong as steel, but also transparent – can you imagine a see-through skyscraper?

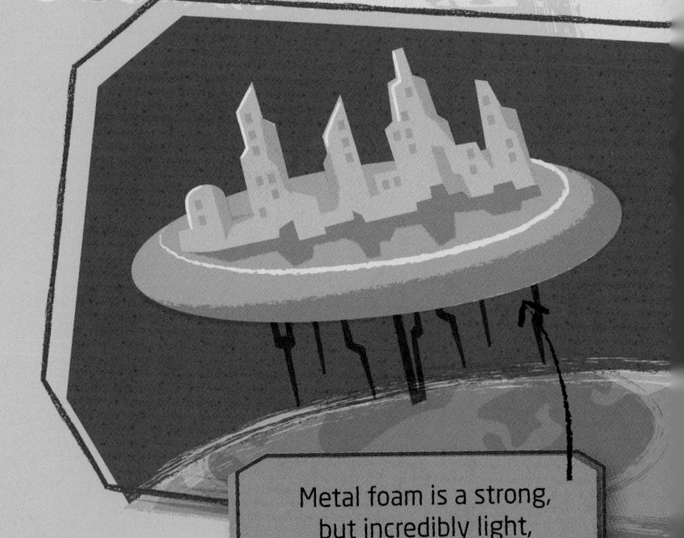

Metal foam is a strong, but incredibly light, material. Perhaps one day it will be used for building cities in space.

GOT IT!

American engineer Stephanie Kwolek (1923–2014) was working on creating new synthetic polymers when she invented Kevlar®. This superlight material was also strong enough to stop bullets! Today Kevlar® is used in bulletproof vests and helmets. It has probably saved thousands of lives.

YOU'RE THE ENGINEER: ENGINEERING STRUCTURES

Find out for yourself how changing the structure of a material can affect its strength and flexibility.

You will need

- Scissors
- A paper cup
- String
- Squares of paper 20 x 20 cm (origami paper would work well)
- Sticky tape
- Small coins
- A notepad and pencil

1. Use the scissors to create a hole in each side of the paper cup. Cut about 30 cm of string and put it through the holes to make a long "handle." Tie a secure knot inside the cup on each side.

2. Take six squares of paper and make them into tight rolls, each with a diameter of about 2.5 cm. Secure each roll with tape.

3. Take another six pieces of paper and stack them on top of each other.

4 Put two flat surfaces of the same height, such as two tables, next to each other, with a small gap between them. Put the six stacked pieces of paper across them so they form a bridge.

5 Hang the cup from the sheets of paper so the string goes directly down the middle. Put coins in the cup one at a time. How many coins can you put in before the paper falls? Write down what you see. Does the paper sag under the weight of the cup alone? Does the paper sag gradually or fold sharply?

6 Now place the six rolled pieces of paper across the gap so that they are right next to each other, with no space in between. Hang the cup again and repeat the experiment. Write down what you see. Which structure holds more coins?

TEST IT!

Try taping the flat pieces of paper together so they cannot slide around on top of one another.

Try taping the tubes together into one long "raft," or three on top of three.

Try the experiment with tubes of different diameters. How does the diameter affect the strength of the structure?

GLOSSARY

absorbent describing things that absorb liquids easily

aluminum a smooth, flexible metal that can be easily shaped

atom the tiny particle that is the basic building block of all matter

biodegrade to be broken down by bacteria and eventually disappear

brittle easily broken

conductor describing something that electricity or heat can travel through easily

consistency how thick or soft something is

fired heated to very high temperatures in a special oven

flexible able to be bent easily

fossil fuel a fuel that comes from the ground, such as coal, oil, and natural gas

inorganic describing things that are made from nonliving matter

insulator describing a material that heat and electricity cannot travel through

kiln a special oven for firing clay to turn it into ceramic

malleable describing something that can be easily molded into different shapes

metal ore rocks that contain large amounts of useful metals

mineral a substance found naturally in the earth

molten describing the liquid form of something that is usually solid, such as glass

opaque not able to be seen through

ore a kind of mineral, such as a rock, that contains more valuable minerals, such as metals

organic describing things that are made from living matter

polymer a man-made material made by creating threads from liquid chemicals

raw material a material in its natural state, before it has been processed

rigid stiff and not easily bent

streamlined having a long, thin, or pointed shape to reduce friction

synthetic describing a man-made material created using chemical processes

versatile describing something that can be used in lots of different ways

INDEX

ANSWER pages 22-23

Step 4: This is known as the "leather" form, so the clay should feel most like the belt.

COLLECT THEM ALL!

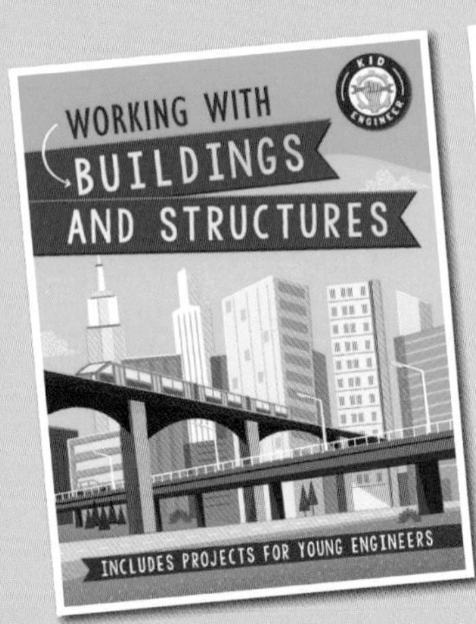